How to Buy Flooring

How to
Buy
Flooring

Doug Liggett

To order additional copies of this book, contact:
Xlibris Corporation
1-888-795-4274
www.Xlibris.com
Orders@Xlibris.com
72488

CONTENTS

INTRODUCTION

In all past ages, there have been traders of flooring: whether it was marble, stone, wood, or rug. Take a moment and consider all the great kingdoms of the past: Egypt, Persia, Rome, Asia, and England. Think of their great halls, palaces, castles, and homes of the wealthy and affluent, and the many types of flooring placed in those great edifices for the kings and magistrates to walk on. In those days, skilled artisans were needed to lay those floors in place. Some were paid for their services, and others did so as slaves or servants.

Today, fine flooring is made available not only to the wealthy like in ages past, but to all classes of people. Skilled professionals and artisans are being used in the floor business of today to satisfy the public's need for flooring. But how do you know you will receive the professional services you deserve from them? What if there was someone who could teach you all about flooring, what to look for in a great flooring store, and how to know you have a seasoned installer? That is where I come in.

Today we live in a fast-paced, busy kind of world. We want the best price and expect top quality all at the same time. We have become used to many choices, especially in the flooring industry. Carpeting, for instance, comes in hundreds of different styles, colors, weaves, textures, and thicknesses. Wood, laminate, ceramic, porcelain, vinyl tile, sheet vinyl, and linoleum also have hundreds of different variations. So how do you know what to buy when you are shopping for flooring? What type of flooring is best in what area? I can help you with these types of questions.

The flooring market is ever changing. New products are developed every year, which adds to the confusion. Some of the flooring you bought just ten years ago may now be obsolete or they have added something new to make the old product better. Or maybe you have never purchased flooring before.

So where do you go to get the real truth about flooring and installation methods? Some say you can get all the answers in the Internet. Many salespeople and store owners say they have the answers. Are they giving you all the facts, or do they just want to sell you the one item that will give them the most commission or make the flooring store the most money? I have been there, and I can give you the unbiased facts.

I started writing this book several years ago when I was a flooring retailer because so many of my customers were unprepared when they came in to shop. Of course I let them take samples of flooring and made an appointment with them to measure their rooms. But I knew that if they were already prepared with measurements and an item from the room or a paint swatch, it would save them from making another trip. I also knew that if they had a basic knowledge about flooring types and where each is best suited in their home, they would feel more comfortable about their purchase.

As a flooring installer since 1982, a flooring service business owner since 1986, and a retail flooring store owner between 2001 and 2006, I have the experience and understanding of how the flooring business works. I know what materials belong where and how it should be installed. I know a good installer when I see one, and I can help you save time and money the next time you purchase any amount of flooring. I can also guarantee the installation will be done right . . . if you follow my directions!

This book was written to help the public save time, get the floor of their dreams at a reasonable price, and know that their installer will be a competent professional every time!

The key is, *you must read this book, follow its direction, and carry it with you when you shop for flooring; then you will save time, money, and get a professional installer every time . . . guaranteed!*

GETTING TO KNOW
FLOORING INSTALLERS

The first thing you need to realize as a consumer buying flooring is that the materials are not a finished floor until they are installed. You can have the best materials money can buy, but if you get a lousy installer, you have wasted your money. Getting a second rate flooring installation is like buying a car that has problems as soon as you drive it off the lot. It may look good at first, but then the carpet starts to bubble or the grout in your ceramic tile job starts to crumble. I hope you see what I mean.

I want to help you avoid these types of problems in the future by helping you recognize a good installer, or one that should not be installing your floor in the first place!

How can I know if I have a good installer?

> I have been a flooring installer for over twenty-five years, and I have seen them come and I have seen them go. There is no real way to know if you have a professional installer every time, but I can point out several attributes of real artisans I have met in the past.

The following is a list of a few of those tendencies.

1. They have a list of references with them at all times and are not afraid to have you call some of them, even before they start on your job.
2. They will call you if they are running late and will give an approximate time when they will be to your job.
3. They will clean up after themselves and will never leave the trash from that cleanup on your property.

4. They stand behind their work for two years, going beyond the usual one-year installation guarantee of the retail store.
5. They never smoke in your house or on your property.
6. They are always courteous and many times will explain what they are going to do before they do it, so you will know what to expect.

Use this list as a guide and give your installer the benefit of the doubt. If they have all the right tools, references, and will give you a guarantee in writing, that will probably be good enough.

Your best option is to use proven installers from a retail store. But if you are buying materials online, let the wholesaler suggest an installer and then interview them before they install for you.

LEARNING ABOUT CARPET

Today's Carpet Choices

In order to select the right carpeting for the right area, it is important to know a little about how carpet is made, the different types of carpet fibers, and what styles of carpet are available. You should also understand how to measure the areas you want covered and how to calculate the square footage or square yardage needed in those rooms. This section will teach you the basics about carpeting, so you won't be fooled by inexperienced flooring salespeople or even the seasoned professional. You may even know more about carpet than many of the flooring salespeople you will meet, and that will be *to your advantage*!

Before we get started, I would like to tell you of one experience I had a few years ago. I went to a house to take up a Berber carpet that had been down only three years (Berber is a popular carpet that has large looped fibers). They had installed the Berber in their living room and den area. It was soiled, flattened where they had been walking, and there were several places where the loops had been pulled and were loose at the seams. They had an inside dog of medium size, and the den had a sliding door to an outside porch.

The lady of the house complained that the store's associate told her that a light-colored Berber carpet would work well in those areas because of its durability and clean ability. She went with the advice of the salesperson. The truth is, a light colored Berber against an outside door with a medium sized dog is the *wrong application*! The Berber loops are made up of hundreds of strands of fiber pushed together in a continuous loop. These fibers are not as easy to clean as you might think. Dirt and oils drive deep into these fibers and are hard to get out. Also, Berber loops are easily pulled by a dog's untrimmed claw.

This is just one example of many in my career as an installer. Putting the wrong carpet in the wrong place under the wrong circumstances is *a tragedy for the one buying the flooring, not for the one selling it!*

Types of Carpet Fibers

In order to understand why certain carpets are best suited in particular areas of your home or office, you must understand what carpets are made of and how they are manufactured.

The majority of the carpet produced in the United States contains one of six pile fibers: nylon, polypropylene (olefin), acrylic, polyester, wool, or cotton. Synthetic fibers make up more than 99 percent of the fiber used by the U.S. carpet industry. Each fiber has strengths and weaknesses that must be recognized. Some fibers have very low resiliency and should only be manufactured in high-density loop pile constructions to limit crushing (pile flattening). Other fibers have the tendency to absorb oily soils and other oil-based compounds (including body oils) and should be carefully considered before installing in areas subject to these contaminants. It should be emphasized that there is no perfect fiber and carpet is a fabric that is subjected to incredible abuse through foot traffic, accidental spills, environmental contaminants, and other abuses.

In general terms, the tighter the twist and the more twists of yarn per square inch of carpet, the longer the carpet will perform before it begins to crush and matt. By taking lots of fine fiber strands (called filaments) and spinning them into a tightly twisted yarn and then locking them all together with heat, you get a much more resilient yarn for making carpet.

Also the height of the carpet's nap is important to the overall performance of the carpet. The higher the yarn stands above the backing of the carpet, the more difficult it is for the manufacturer to make the carpet crush and matt resistant. So the shorter the pile height and more dense the yarn, the more crush resistant and longer the like-new appearance of your carpet will last.

If you need more facts about carpet and how it's made, go to *www.floorfacts.com* and get more answers.

If you want to be a good carpet shopper, you should know the advantages and limitations of each type of carpet fiber. In many cases, you will know more about carpet fiber than the salesperson attending you. Please, take the time to read this section before shopping for carpet.

The following explanation of each type of fiber will help in your purchase of carpeting. It will help you identify which fiber will perform the best under the conditions that will prevail in your room. You, the buyer, will finally be in control of your next carpet purchase!

Wool Fiber

Wool is a natural product derived from the shearing of sheep. Most fabrics use long and short fibers interlocked together, which makes it nature's staple yarn. It has the ability to hold or release water vapor as required by climate condition, making it a natural humidifier. It keeps dust and dirt near the surface of the carpet pile where it can be released by vacuuming. It is naturally flame resistant because it chars instead of melts.

This fiber has a natural *resilience* (bounces back when under a lot of traffic), resists abrasion, retains its *texture* (feels the same after many years), is resistant to soil, and is very cleanable. It will do well in any inside dry high-traffic area you would like to put it in. However, it is expensive, has no static-free tendencies, holds on to stains, and some people are allergic to its fibers.

Nylon Fiber

Nylon is a synthetic fiber first made into carpet in the 1940s. It is known for its strength and abrasion resistance. It is resilient, has good texture retention, and is very cleanable.

This fiber is long wearing, resilient, comes in many vibrant colors, and is inexpensive to produce. It is also a bulky fiber that covers well.

Nylon carpet fiber, in today's many variations, is the most popular in the world of carpet fiber. I believe it is a very good choice, especially the advanced nylon products.

However, the cleaning of this fiber should be done with strict adherence to manufacturers specifications.

*One thing to note: Yarns can be either *bulked continuous filament* (BCF) or *staple*. Polymer is forced through a spinneret (extrusion) in uninterrupted filaments, which are then formed into a *bulked continuous filament yarn*. These fibers may also be *chopped into short fibers and then spun into staple yarn*, twisted, and set with heat to hold the twist. A tighter twist is more important in cut pile because it resists the ends of the yarn from untwisting and matting together during wear and cleanings.

If you are looking to install nylon carpets, make sure the fiber is *BCF* nylon. If you see on the back of a carpet sample a label that says "CFN", or "BCF," that means this carpet is made from continuous filament nylon. *Continuous filament nylon starts out as one long continuous strand before made into yarn.* The other type is called *staple* nylon, which means *each fine filament starts out as a short strand about 7 inches long before it is spun into yarn* and then tufted into carpet. As the short, unsecured pieces of staple nylon come loose from the yarn pile you get pilling on the surface of the carpet (lots of yarn fibers coming off the carpet). This happens very little with a continuous nylon carpet.

Remember to go to *www.floorfacts.com* and read more about this type of carpet fiber.

Acrylic Fiber

This fiber was introduced just ten years after nylon and has many wool-like properties. It is a bulky fiber often blended with agents that guard against potential flammability. Even these improvements haven't given this fiber the reduced flammability of wool.

This fiber is warm, has the appearance of wool, resilient, resistant to abrasion, fine texture retention, and has good cleanability.

As with nylon, acrylic fibers must be cleaned according to manufacturer's specifications.

Olefin (Polypropylene) Fiber

This fiber is the lightest of commercial carpet fiber and was produced around 1960. Virtually nonabsorbent to liquids, this fiber cannot be dyed, but the color must be placed in the fiber when it is processed. It is often entangled with nylon to add flecks of color. Indoor/outdoor carpets are often made of this fiber.

Olefin is resistant to abrasion, very cleanable with some resilience and texture retention. It is inexpensive, moisture resistant, fade resistant, and stain resistant, with low static features. However, it does not hold up very well in heavy foot traffic, it has a limited color range, its flammability rating is lower than nylon, and it has a low melting point (high friction burns are possible in this product).

Polyester Fiber

Polyester was produced after 1965 and has become popular because of its soft, luxurious appearance. It is the most produced and used man-made fiber in the world. It is the most recyclable fiber produced in bright lustrous shades.

This fiber has a good resistance to abrasion, and when fibers are adequately *heat-set* (twisted fibers are heated to bond them together), they are resilient and texture is retained. It is naturally stain resistant and has the highest melting point of any carpet fiber, but it is less durable than nylon.

In my opinion, the nylons are the most resilient to high traffic. Unless the polyester is mixed with nylon, this fiber is not meant for high traffic areas.

The Next Generation of Fiber

Around the year 1995, an enhanced polyester polymer called PTT was developed with the resiliency and durability of nylon. This polymer fiber is as dye able, durable, and resilient as nylon with the added features of permanent stain resistance, color clarity, and softness.

This fiber resists all acidic stains found in food and beverages. Foot traffic will not wear the stain resistance off, and recommended cleanings will not weaken it. It withstands bleaching agents like chlorine and even acne medications. Oil and wax-based stains can be cleaned as well. These fibers do not absorb liquids, so they easily resist and release stains. PTT polyester fiber cleans better than a "topically treated for stain and soil" nylon.

In my opinion, and the opinion of most fiber experts, this carpet fiber is superior to most any other and belongs in most areas of your home or office no matter what conditions are present!

The Making of Carpet

How was carpet made anciently?

Woven carpet rugs have been made by hand for thousands of years and are still made by hand today in many Old World countries. The weaving of pile rugs is a difficult and tedious process that, depending on the quality and size of the rug, may take anywhere from a few months to several years to complete.

Specific ancient tools were used to make it strong and resilient. In order to operate the loom, the weaver needs a number of essential tools: a knife for cutting the yarn as the knots are tied, a comblike instrument for packing down the wefts, and a pair of shears for trimming the pile. In Tabriz the knife is combined with a hook to tie the knots that lets the weavers produce very fine rugs as their fingers alone are too thick to do the job.

Today woven carpet is created on looms by simultaneously interlacing face yarns and backing yarns into a complete product, thereby eliminating the need for a secondary backing. A small amount of latex-back coat is usually applied for bulk.

These facts are found in the Mohawk University Workbook, Mohawk Industries Inc.

How is carpeting made today?

The look and performance of a particular carpet is determined by its construction. In corridors, lobbies, offices, classrooms, hotel rooms, patient care facilities, and other public areas, looped piles of short, dense construction tend to retain their appearance and resiliency, providing a better surface for the rolling traffic of wheelchairs or food carts. Cut pile or cut-and-loop pile carpet are very good choices for administration areas, libraries, individual offices, and boardrooms.

Understanding carpet construction will help you choose the carpet that will provide the best performance in a particular location of your home or office. Commercial carpet is primarily manufactured by tufting or weaving. Each process will produce quality floor coverings, but tufted carpet accounts for 95 percent of all carpet construction. Both tufting and woven manufacturing are efficient and employ advanced technologically to provide the capacity for a myriad of patterns and floor covering.

Make sure and go to *www.floorfacts.com* to learn more about carpet construction.

What is tufting?

Tufting is a method of fabricating carpet begun in the early 1950s. Since the late 1970s tufting has increasingly become more sophisticated and computer design-driven. The tufting process begins when measured lengths of yarn are fed off beams or creels to become the carpets' surface as they are sewn through a primary backing material.

The yarn is fed into needles through tubes that go from the creel racks to the tufting machine. The needles of the tufting machine's needle bar form loops that are hooked by loopers on the underside of the backing material, which remain level looped or textured loop carpet, or are cut with knives to create cut pile or cut/loop pile carpet. Now the tufted fabric is ready for coating and finishing.

After yarn is sewn through a primary backing by a needle (as is mentioned above), it is then adhered to a secondary backing material with latex adhesive. A computer controls the amount of latex, and the amount depends upon the weight of the carpet. Too little latex will cause delamination, and too much will cause stiff and brittle carpet.

Make sure and access the Internet and google "Floor Buyer's Guide" to get more formation about carpet and many other flooring products available to consumers.

The important thing to remember is that the amount of latex and the secondary backing give the carpet stretch and hold (which is very important in the installation process). A proper stretch means no ripples in the carpet for the life of the carpet. It means the carpet will be easy to clean and maintain. It means that it will be comfortable to walk on and the fibers will bounce back even after years of use. When the carpet fabric is made properly and installed properly, you the owner benefit the most. You won't have to replace it until you're tired of it!

Styles of Carpet

There are basically seven different styles of carpet: textured, saxony/plush, Berber cut pile, shag, cut/loop, Berber loop, and multi-level loop. Each has a different texture or look to them. The following explanations of each style will help you identify what the carpet pile looks like and what advantage or disadvantage each has.

What is a texture?

A *texture* is one of the commercial design carpets that have colorful patterns and are used to promote a theme in the room it is laid. This type of carpet can withstand heavy traffic because it has a short, dense pile construction and the designs in the carpet break up wear patterns. It is mostly used in commercial applications like hotels and motels and can also be used in kitchens at home.

What is a saxony/plush?

One of today's more popular styles of carpet is called plush, or saxony. Its twisted fibers are cut evenly, and it feels smooth to the touch. This is called the "hand" or how the carpet feels through the touch of your hand. There are several different variations of plush carpeting.

The general rule is, the denser or thicker the carpet is, the more it stands up to traffic. When you buy a plush or saxony carpet for a high-traffic area like a living room or den, it should be *dense* (the closeness of the fibers to each other), with a shorter *pile* (the height of the fibers from the backing to the top of the fiber), so that it will hold up to lots of foot traffic. So if you are buying a plush carpet for say a bedroom it does not have to be as dense because there is not as much traffic.

However, even a short pile dense plush is not always the right way to go. A light-colored dense plush can collect oils from bare feet, animals, and kitchen floors showing walk patterns over time. Frequent cleaning will be needed to keep the area looking new.

What is a Berber cut pile?

I am proud to say a new development in carpet manufacturing has brought back the wear ability of the old shag. It is a cross between a Berber and a plush carpet and can be called either a Berber cut, frieze, or twist. It has the softness of a plush and the wear ability and strength of a Berber. The fibers are twisted together and heat-set like a plush, and then the multi-fiber twist are twisted again like a spring. This New Berber has a soft feel, comes in many more colors than regular Berber, and has the wear ability of Berber without the difficulty of cleaning.

I highly recommend this type of carpeting in most areas of your home. For a frieze, denser is not always better. You don't have to go to the expense of a dense twist (Cut Berber) carpet because the twisted fibers spring back no matter how thick or dense the fibers. This is where technology has done its magic, bringing together the better attributes of the two products and making one!

Again I go to my experience as an installer. When this carpet is installed, the seams are nearly invisible. Rarely is there a *discoloration problem* (when the carpets are seamed together it looks like different shades of carpet), and it installs easily with just the right amount of stretch ability. In my book, it is the carpet of choice for most areas you might want carpeting in. It is what I urged my customers to buy when I sold carpet at my store.

What is shag?

Many of us remember the old shag carpet that became popular in the '60s. The twisted yarn was thick and many times two to three inches in length. It lay on the floor like hair on a person's head. You could even brush it with a shag rake, and it was usually made of bright colors.

Today the old shag has a new look. Instead of only thick, long twists throughout the carpet, it now has variety. Some twists are short, and some are long; some are thick, and some are thin. It even has loops throughout the carpet to help give the carpet bulk, helping it stand up to heavy traffic.

The "New Shag" would be well placed in a study, bedroom, living room, or den. It is expensive, but it will look good for long periods of time no matter where you install it.

What is a cut/loop?

The sister to the plush is called sculptured or *cut/loop*, noted for its looped design woven amongst the pile of plush fibers. It rarely comes in one shade, is multi-colored in design, and it comes in a variety of pile heights and densities. It is very popular with those who have less to spend in high-traffic areas because it does not show wear patterns as easily as a regular plush with lower density. You will find this type of carpet in apartments, rental property, and bedrooms much of the time.

The only drawback to installing the sculptured carpet in a high traffic area is that over time, this type of carpeting will *matt* (flatten), depending on the amount of foot traffic or animals in the house. If you are buying a sculptured carpet for say the living room, I would suggest a denser product with a lower pile.

What is a Berber loop?

Another type carpet that made its debut in the '70s is the *Berber loop*, known for its large continuous loops. It has been popular for years but is losing its popularity. Mainly because it is hard to clean, the loops are easily pulled, and it collects oils and grime. Another complaint is that the fiber is rough to the touch and comes in colors that are too plain. However, it is perfect in high-traffic areas, including stairs. It rarely shows wear patterns and is reasonably priced, so it fits most budgets.

What is a multi-level loop?

Multi-level loops were created to handle extreme traffic and can be put on porches and in game rooms. It has a very small tight loop and is mostly used in business offices and leased property.

Like the large Berber loop, this type of carpet is rough to the touch, and the loops can be easily pulled out of the carpet backing by animals or furniture like desks or cabinets. However it rarely shows signs of wear and is inexpensive to install.

*From level loops to Berbers, there is an important factor to look for when buying a crush-resistant looped carpet. This is the density of the loop and how much free space is underneath the loop. Like all carpets, the more filaments that are packed into the yarn and the tighter the loops, the better the performance will be.

So let's review:

Textured carpet has primarily a short, level, tight pile usually dense in construction and has a rough feel to it. It is mostly used in commercial applications such as hotels or restaurants.

Saxony/plush carpet has a level pile and is soft to the touch. The tighter the pile is packed, the heavier traffic it will handle. It comes in an array of different thicknesses and colors.

Berber cut pile carpet or twists (friezes) have the softness of a plush, but the durability of a Berber. They come in an array of thicknesses and colors, even brilliant colors. This product does not have to be dense to wear well.

Shag carpet in its new construction is made up of advanced fibers that are just as stain resistant as before, but with greater resiliency. The twisted heat set shafts of fibers are long, like the old shag carpet but with some new textures woven in.

Cut/loop or sculptured carpet has a level pile with loops running throughout, showing designs in the pile. It is not made with a very tight pile because the design breaks up wear patterns. It comes in an array of multiple colors, patterns, and thicknesses.

Berber loop carpet is a heavy looped carpet that many times has a design pattern built in for effect. The pattern in Berber must be matched for proper installation and requires more material to install. The large loops give it durability but also make it harder to clean. The fibers used in its construction make brilliant colors unavailable to Berbers.

Multi-level loop carpet (which many refer to as indoor/outdoor) is a small, tight-looped, multi-level carpet that comes in multiple shades of color and are best in high-traffic areas.

Choices of Carpet Cushion

Now that we have discussed fiber manufacturing, carpet construction, and types of carpet, it is only appropriate to talk about the carpet cushion or padding. Pad is placed under carpets when a soft feel underfoot is wanted, after which the carpet is stretched into place. In order to properly stretch carpet, an installer must have strips to hold the carpet in place along the walls. This product today is called tackstrip.

Many years ago, when an installer stretched carpet into place, he sewed the seams by hand and used tack nails to hold the carpet's stretch. Today, the seams are glued together with seam tape and tackstrip is used to hold the stretch against the walls. Padding is placed in the middle of the rooms to the edge of the strips to protect the carpet fibers from being crushed, and to give the customer a soft walk.

That being said, pad is used to protect the carpet fibers and give comfort. It is important to choose a cushion that is appropriate for the particular end use. The best pad for the area will protect the carpet from being crushed by foot traffic and at the same time give a soft walk. A good quality carpet cushion will extend the life of any carpet, regardless of the quality of the carpet. Carpet cushion serves a vital purpose, which is often over looked. Good quality carpet cushion acts as both a shock absorber and a spring that helps improve a carpet's overall performance. Always review the carpet manufacturer's cushion recommendations prior to purchasing a carpet.

Families with pets or those worried about spills should look for carpet cushions that are made specially to block spills from penetrating the cushion and are odor resistant.

What is carpet cushion?

Carpet cushion or padding is the material used to protect the carpet fibers from being crushed by foot traffic. It is placed under the carpeting and bonded to the floor. It is either glued or stapled depending on the floor type, which keeps it from moving while the carpet is being stretched into place. Each variety also has grades, which vary by weight or density. Density is one of the most important factors in buying the right cushion. All types of cushion can be made lighter or denser, depending on the process used to manufacture it. There are three basic types of cushion: foam, rubber, and fiber.

How is each of these three types of cushion made?

Foam is one of the three types of cushion mentioned above. Foam comes in three varieties: prime, bonded, and froth.

Prime polyurethane foam is a firmer version of the foam used in furniture, mattresses, and automobile seats. This pad has no hard spots, and the air pockets within the foam give a softer feel underfoot. Because of the numbers of air pockets, it has a tendency to flatten over time. The denser the foam pad, the more expensive it is and the longer it takes to flatten. It is not recommended for installation under Berbers and its standard width is six feet.

Bonded polyurethane foam (sometimes called rebond) is formed by combining chopped and shredded pieces of foam, in different sizes and usually different colors, into one solid piece. Liquid foam is used as an adhesive to hold it together. It usually has a surface net to make installation easier and improve performance. It is the most popular cushion used residentially. Rebond cushion is measured by the density and thickness of the cushion. Density is calculated by weighing one cubic foot on a scale. If the cube weighs six pounds, the overall density is six pounds. This type of cushion can be purchased by weight and by thickness. Usually the denser the material, the more it costs. However, the denser the material is, the longer it will last.

Froth polyurethane foam is made with carpet backing machinery. Liquid ingredients are applied to the back of the carpet or to a nonwoven material to attach the foam directly to the carpet. This (carpet/foam cushion) combination is especially popular with those who want to install the carpet themselves.

Rubber comes in two basic types of sponge rubber carpet cushion: waffled and flat.

Waffled rubber cushion is made by molding natural or synthetic rubber. Heat cures the rubber and forms a waffle pattern. This produces a soft, resilient cushion perfect for residences. Because of its weight, it comes in 41/2 foot widths.

Flat sponge rubber is a firm, dense cushion that has a flat surface and is usually used when Berber carpet is installed. This cushion is also used when a carpet is to be glued to pad. It comes in different weights and thicknesses and is measured by weight in ounces per square yard of material.

Fiber cushion uses existing fibers (either natural or man-made) that are interlocked into a sheet of felt. Fiber is usually recommended under Berbers and has a very firm feel underfoot. There are two distinct varieties of fiber cushion: natural and synthetic.

Natural fibers include felt, animal hair, and jute (the material used to make rope). This is one of the oldest types of carpet cushion dating back to the earliest days of machine-made carpet.

Synthetic fibers include nylon, polyester, polypropylene, and acrylics that are needle punched into dense cushions that have a firm feel. It can be made in any weight or thickness.

Knowing a little about carpet cushion will help you decide which type of cushion you should place in what area in your home or office. The following is a chart that meets minimum guidelines from the Carpet Cushion Council.

Minimum Recommendations for Residential Applications

Class A—Light and moderate traffic
Class B—Heavy-duty traffic
Prime Polyurethane Foam—¼ inch / 2.7 pound cushion for both classes
Bonded Polyurethane Foam—3/8 inch / 5 pound cushion for **Class A**
                            3/8 inch / 6.5 pound cushion for **Class B**
Froth Polyurethane Foam—¼ inch / 10 pound cushion for **Class A**
                            ¼ inch / 12 pound cushion for **Class B**
Waffled Rubber Cushion—3/8 inch / 48 ounce / 14 pound for **Class A**
                            3/8 inch / 64 ounce / 16 pound for **Class B**
Flat Rubber Cushion—¼ inch / 56 ounce / 18 pound for **Class A**
                            ¼ inch / 64 ounce / 21 pound for **Class B**
Natural Fiber Cushion—¼ inch / 40 ounce / 12.3 pound for **Class A**
                            3/8 inch / 50 ounce / 11.1 pound for **Class B**
Synthetic Fiber Cushion—¼ inch / 22 ounce / 6.5 pound for **Class A**
                            3/8 inch / 28 ounce / 6.5 pound for **Class B**

*Use these figures in conjunction with the section "Styles of Carpet" to pick the type of cushion that is recommended above for that particular style of carpet.

Example: If you pick a Berber carpet to be installed in a high traffic area (class B). You should pick a Flat Rubber Cushion or a Natural/Synthetic Fiber Cushion (for class B) to be your padding.

*This chart illustrates products that meet minimum guidelines. Better grades of carpet cushion than the minimum suggested are always recommended when possible to provide more support and cushion for carpet.

The Mohawk University Workbook has many more facts about carpet cushion. To get more information, go to a Mohawk Flooring dealer near you or google Mohawk Industries Inc.

Types of Carpet Installation

How is carpet installed?

There are really only two ways to install carpeting. One is called the stretch-in method, and the other is the direct glue-down method. Both have their place in the installation of carpeting. But it is important to know which method is best suited in what area to get the most out of your carpet purchase.

With the stretch-in method carpet is stretched onto tackstrip, which are placed along the edges of a room about 3/8 of an inch from the wall on all sides (depending on the carpet used). Between the strips, placed in the middle of the room, the installer will have secured the carpet padding. The tackstrip is used to hold the carpet in place as it is stretched into the room by a knee kicker and carpet stretcher (power stretcher).

The *knee kicker* is used to secure the backing of the carpet to tackstrip, which is placed along the wall of the room. This tool has a head with adjustable spikes and a kneepad to allow an installer to kick into it with his knee without being hurt. As the head of the knee kicker grabs the carpet, the installer uses his knee to force the carpet backing onto the small metal spikes set at angles toward the wall in the tackstrip. Those metal spikes hold the carpet in place.

Once the carpet is secured to the tackstrip along one wall, it's time for the *power stretcher.* This tool is a larger version of the knee kicker with a few differences. It has a foot bar that is used to push against one wall in order to stretch toward another. It also has adjustable tubes that are placed together to go the length or width of the room. These tubes connect to the foot and head of the power stretcher. The head of the tool has adjustable spikes set at angles to penetrate the carpet, and with its pistonlike lever, it grabs and stretches the carpet so it can be placed on the spikes of the tackstrip. Once the carpet is secured along one wall and power stretched to the other, it is now tight like the head of a drum.

These are the carpet installer's stretching tools. The knee kicker sets the carpet on the tackstrip, and the power stretcher stretches the carpet properly in place.

With the "direct glue-down" method, there are no tackstrips. The carpet is placed on the floor without carpet cushion. An adhesive is applied to the floor, and a seventy-five-pound carpet roller is used to press the carpet into the adhesive. After which the excess carpet is cut away from the walls and tucked into the corner, as if the walls were built on top of the carpet.

How is a carpet seam made?

At this point I should instruct you on how a seam is made both as a stretch-in and a direct glue-down method. You will need to have this information when picking a place where the seam should go in your room or if you want to make a seam yourself.

When a carpet is manufactured, loops of yarn are stitched into the carpet's backing. Not only are they sown at an angle but also in rows. These rows are used by the installer to put together a seam, especially in looped carpet such as Berber loop and multi-level loop carpet.

To set up a seam, an installer needs to lay the carpet out the way it should lay in the room. That is, it should fill the room, going up the walls slightly on all sides, and overlapping where the seam should be.

For instance, if the room is thirteen feet six inches wide and fifteen feet six inches long, and if you are laying the seam on the backside of the length of the room, then you will have one drop of carpet twelve by thirteen feet nine inches and the other four by thirteen feet nine inches (or pieces that will fit that area running in the same direction).

If you are stretching the carpet over padding, the seam will be put together by applying seam tape under where the carpets meet at the seam. This is done after the carpet is cut along the row of yarn (for looped carpet) or straight cut (in the case of a plush or Cut Berber carpet). The adhesive on top of the seaming tape is melted by what is known as a seam iron as the carpet backing is put together behind the hot iron by the installer. The seam is then pressed into the hot adhesive with a carpet tractor and left alone until cooled.

If you are laying the carpet as a direct glue-down, seam tape is not necessary. However, a seam sealer will be required as you put the seam together while the carpet is being pressed into the adhesive. A carpet tractor is also used in this installation method to make sure the seam is securely placed before the adhesive dries.

In either case the seam must meet both sides of the carpet in a straight line. The installer is trying to place the backing of the carpet together as if a cut was never made. For the women or men who like to sew, this means the material must meet perfectly and run in the same direction, when sown together.

Which type of installation method is the best?

This depends on where you are installing carpet, because the environment determines the mode of installation. You should determine the use of the room, the type of subfloor you have in that room, how you want the carpet to feel underfoot, and your budget.

For example, if you have decided to put carpet in your playroom, and there's a pool table in the middle of the room, you should probably place a Berber or multi-level loop carpet on the floor using the direct glue-down method of installation.

In order to select the best installation method, you need to consider the advantages of the following types of installation.

Stretch-in advantages:

- Feels more comfortable than direct glue-down
- Extends the carpet life
- Minimizes matting and pile crushing
- Increases insulation value
- Intensifies the vacuum's removal of dirt and soil
- Minimizes any subfloor problems
- Reduces the cost of later removing installed carpet
- Absorbs more sound
- Allows better matching of patterned carpets

Direct glue-down advantages:

- Eliminates the cost of the cushion
- Requires less labor
- Minimizes carpet buckling when temperatures are extreme
- Eliminates the need for any restretch
- Seams are more durable
- Eliminates peeking seams

- Installed in rooms of any size
- Allows easy access to electrical or telephone lines under the floor

Advantages in cases using carpets with attached cushion:

- Improves appearance retention and foot comfort
- Acts as an effective moisture barrier
- Improves thermal and acoustical contribution
- Eliminates the second adhesive required for double glue-down installs

How do I know I have a good carpet installer?

Purchasing carpet for your home is a big decision. Making sure your new carpet is installed properly is just as important as choosing the right style, color, and type of carpet. Knowing the facts about your installer will help you get better results.

Many carpet dealers have professional installers and will handle the details of installation for you. *You may save money by purchasing your carpet from a source that does not provide installation services, but if you choose this route, be sure to hire a professional installer who knows what he is doing.*

There are several ways to find out if you have a good carpet installer;

First, ask about the length of time the installer has been installing carpet and make sure his experience is in residential or custom homes. Five years is an acceptable amount of experience in most cases.

Then, consider the complexity of your job, because the more complex you feel the job will be, the more experience you want to require of your installer.

Also, make sure your installer is licensed and bonded. To be certain, ask for copies of licensing and insurance information.

You should, then, ask for several references and then follow up. Depending on the size of your job, you should ask for three or more references to get the right recommendations. Request recommendations for installation jobs performed at least six months prior so you can find out if any problems have arisen since the job was completed.

For added peace of mind, be sure to find out what kind of installation warranty the installer provides. Should problems arise in the future, you need to be certain about the terms of your warranty.

Once you feel confident that you've chosen a good installer, be sure the contract you sign includes all materials, labor, and anticipated start and completion dates for the project.

Remember, good carpet installers have a power stretcher and will use it even if it is in a small hallway. They will stretch carpet away from seamed doorways to the far wall, to keep carpet from rolling at the door in the future.

Finally, good installers clean up after themselves and will even offer to vacuum after the job is done.

If you purchase your carpet from a dealer who handles installation, make sure all the areas mentioned above are covered thoroughly. Peace of mind is important when you're spending money to put new carpet in your home, whether the job is large or small.

LEARNING ABOUT HARDWOOD

Today's Hardwood Choices

Years ago, houses were mainly built on block or brick with a basement or crawl space. The living space was covered with solid wood plank floors in the living rooms and bedrooms, and linoleum in the kitchen and bath areas. After a few cold winters, these floors were eventually covered with carpeting.

But today we are taking back those "wood floor days" with a new generation of hardwoods made to last. This new generate is called engineered hardwood or laminated wood floors. It is made in layers like plywood, but the top is made to last longer without dents or scratches than regular hardwood. Many even have a warranty stating how many years the flooring should look good under normal conditions.

Types of Hardwoods and Their Construction

The manufacturers of hardwood flooring make only two types: solid wood and engineered. All hardwoods come in different thicknesses, stains, and warranties. Many of today's hardwoods are made tongue and groove, which fit together on all sides. These wood planks come in many different widths and lengths depending on the style of plank and the manufacturer. They come this way in order to help an installer stage placement of planks on your floor so joints do not meet. It is important that you understand the different types of hardwoods and their uses when buying hardwood floors.

What is a traditional hardwood or solid wood plank?

Let's start with the traditional type of hardwood that is now sold by retailers. It's not engineered like plywood but is a solid piece of wood, stained to bring out the wood's natural beauty and protected by a urethane coating. It comes in several thicknesses up to three-fourth-inch thick. All solid hardwood products must be stapled down during the installation process.

In order to install this type of floor on a concrete slab, a subfloor of three-fourth-inch plywood must be laid before the solid wood plank is installed. This can be a real drawback even if money is no object because a three-fourth-inch plywood subfloor and a three-fourth-inch solid wood plank make this floor set one-and-one-half inches above the concrete. This height can make it hard for other floors to meet up to it properly. However, wood transition pieces can help.

What is an engineered hardwood?

Engineered wood floors also referred to as laminated wood floors are often confused with laminate floors by consumers. Engineered wood floors are *not* laminate floors, so be careful when shopping for a new floor. Engineered wood floors are constructed differently from solid wood floors and offer some advantages over solid wood floors.

Thanks to advancements in manufacturing technology, engineered wood floors can be used in almost any room in the home. This includes installing over concrete slabs and some types of existing flooring. Homeowners can now enjoy the beauty of a real hardwood floor in areas that were excluded because of the drawbacks of solid wood flooring.

Most engineered wood floors are prefinished at the factory. Prefinished wood floors are ready to be walked on right after the installation is completed. Most factory applied finishes are UV-cured with ultraviolet lights, creating a much harder finish than a job-site finish. Also, the manufacturer can apply more coats of finish, thus giving added protection.

The top layer of an engineered wood floor is available in three variations: rotary peeled veneers, sliced, and sawn face.

Rotary Peeled Veneers—logs are processed in a conditioning vat and put onto a large wood lathe. The wood veneers are then peeled off the logs in long strips. This gives maximum yield from the log.

Sliced—lumber is prepared in a conditioning vat, and the lumber is then sliced. The lumber is first cut from the log in a sawmill, and then it is processed for slicing.

Sawn Face—a traditional process where lumber comes from the log in a sawmill. The lumber is graded and sorted for maximum return. The lumber is then sewn into the desired thickness making it ready for engineered construction.

Which top layer is better?

Each top layer variation has its good and bad side. For example, *rotary peeled* uses the most raw materials for the lowest cost. It is also the least visually appealing with a weak grain structure. The *sliced* has a yield with medium cost, is better looking, and is more structurally sound. *Sawn face,* however, has the highest cost using the most raw materials, is the best looking overall, and has the most detailed grain structural because of the process used to get to the grain.

How is engineered hardwood constructed?

Wood sections of ply are stacked on top of each other in opposite directions during the manufacturing process (like plywood). Reversing the direction of each ply as it is placed on top of each other counteracts the wood's tendency to expand and contract when moisture is present. This manufacturing process makes the planks much more dimensionally stable than solid wood planks. This stabilizing effect allows engineered wood planks to be installed over concrete floors.

The top layer (the stained ply you see) of an engineered wood plank is often a different wood specie than the other plies under it. Because of this, manufacturers are able to offer a wider variety of domestic and exotic hardwood species while keeping the costs down. Engineered wood floors come either in a three-ply or a five-ply plank.

Where can you install engineered hardwood?

Engineered wood floors can be used over all grades of subfloor, including below grade concrete slabs.

The planks are generally stapled-down, glued-down, or floated over different types of subfloors, such as a wood or concrete. The glue-down and floating installation methods allow the engineered wood planks to be used over concrete slabs, since you cannot staple into concrete. Also, some engineered wood floors can be floated over an existing floor, such as tile or vinyl flooring. Be certain the old floors are down tight in all places before you install over them.

Engineered planks come in a variety of widths, including 2 and 1/4", 3", 5", and 7" widths. Many plank styles can even be mixed, such as three-five-seven-inch planks installed side by side. Varying the board widths changes the total appearance of the floor.

If you need more information about laminated wood, go to *www. floorfacts.com* and look under the subtopic "engineered wood."

Now that you know the difference between solid wood and engineered hardwood, it will be easier making a choice on where each type should go. Be warned that solid wood planks do not belong in wet areas like kitchens, bathrooms, or utility areas. However, entry areas are fine because they see a limited amount of moisture. Just be sure to wipe up water as soon as possible to keep water from going where hardwood is sealed together.

Installing Hardwoods

Installers that are certified to install wood flooring have spent hundreds of dollars on specific tools that are only used in hardwood installation. However, unless you have done this kind of work in the past and have access to all the right tools, I would suggest you leave this one to the professionals.

What tools are needed if I install the hardwood myself?

The following is a list of the main tools you will need and how they are used:

Hardwood plank nailer—hammer stapler or air driven stapler used to drive brads into the tongue of the plank to secure to a wood subfloor. (**Solid wood plank must be stapled** but engineered can be glued or stapled.)

Heavy-duty jig saw—used to cut away notches for vents, plumbing, or electrical.

Radial arm saw—used to make clean cuts from the top of the plank and angle cuts when needed.

Small table saw—used to make length of plank cuts when needed.

Notched trowel—used to spread adhesive in the proper amounts to concrete or wood subfloor for the installing of engineered wood plank.

How is hardwood installed?

The key to starting any installation of wood is to figure out which direction you want to run your planks. To make a room look longer, you must run your planks in the direction of the length of the room (long point to long point). It is up to you which way your planks run. A general rule is to run the planks the length of the room.

If you are using adhesive to install your wood floor the length of the room, you should begin by making a straight line down the middle of your room, perpendicular with the wall. Trowel out the adhesive from the line in the middle to the wall (glue out three feet at a time for best results). Place the edge of your planks along that line, finish that row, then go to the next, staggering the planks until you get to the wall.

If you are nailing a solid wood floor, you should place your first row of planks at the edge of your wall to your right (or left if it is better), with the tongue of the board pointing into the room. The plank stapler shoots brads at an angle into the tongue of the board, securing it to the floor. The first three rows will not be nailed by the stapler because of its size and angle. Loose lay or glue those first three rows putting a three-eighth-inch shim along the wall to hold them in place while nailing the fourth row with the stapler. Continue this action until you have made it across the room.

Wood products naturally expand and contract with the amount of moisture in the air. So it stands to reason that the ends or the sides of the wood plank should not be pressed against the wall or a stationary object. Make sure there is a one-fourth to three-eighth inch gap between the plank and the wall or object when installing solid wood or engineered wood flooring.

When in doubt, read the manufacturer's installation directions, which are usually in each carton of wood flooring.

What if I want to install the hardwood floor myself?

If you are thinking about installing your own wood floor, listen to this.

Imagine crawling on your hands and knees on a hard floor six to eight hours at a time. Wood planks are all tongue and groove, and you must force them together tightly one at a time, both on the sides and on the ends to install them properly.

When stapling a solid wood or engineered wood plank to a wood subfloor, if you don't have all of the needed tools (as stated above), you will need to rent or buy them before beginning the job.

The adhesive can get on the tips of your fingers as you press each plank into place, while gluing an engineered plank floor to your subfloor. This can end up on top of the wood. It must then be wiped off before it dries, so you must clean while you install.

It is also recommended that you stagger the joints (where the planks meet end to end), and they must not meet anywhere on the job. If you lay a bad piece by mistake, you must remove it from the adhesive or pull the stapled piece from the floor and start again, which can be very difficult. This is not counting all the different tools you need to complete the job. I hope you get the picture.

If you decide to let a professional install your wood, make sure he will do the job right.

How do I know if I have a professional wood installer?

References are crucial when hiring someone to install your wood floor. Look for someone with at least five years experience. Call at least three of those references before they start the job. If they check out, you probably can count on their expertise.

A professional wood installer will deliver or have the store deliver your cartons of wood plank two to three days before the day of installation, so the wood can acclimate.

A good installer will keep the area clean and organized during the installation and will clean and dispose of debris after the job is complete.

They will be insured and bonded and will also have all the tools necessary to complete your job from start to finish.

And finally, they should be courteous and considerate of your feelings at all times, which could include not smoking on your property when you don't smoke, not using bad language, and things of that kind.

LEARNING ABOUT LAMINATE

Today's Laminate Choices

There are many new laminate flooring styles and colors available today. Changes in laminate technology have improved the designs, the realistic look, and plank shapes of laminate flooring. Laminate flooring is now offered in wood plank designs with many manufacturers offering the looks and colors of stone, slate, and tile designs with grout lines.

Laminate flooring is not to be confused with laminate wood floors. Laminate wood flooring is constructed completely from wood plies laminated together while laminate floors consist of some type of fiber core interior, with a printed top layer, and a coating of melamine.

Laminate flooring has many advantages. They can give a realistic look of ceramic or stone tile without having to clean soiled grout joints and also save on installation charges. If you have pets or you're concerned with tearing up your old floor, a floating laminate floor may be your answer. A floating floor means the individual planks are locked together but are never attached to the subfloor underneath. These floors have a tongue and groove locking system, which tightly joins the planks and tiles together.

Laminate floors come in either planks or square tiles. Top manufacturers of today's laminate flooring are able to create a very realistic wood plank and tile look. The melamine finish wear layer is easy to care for and very durable. This finish layer rejects most stains, spills, burns along with being scratch resistant.

How Laminate Flooring Is Manufactured

Laminate floors are actually several different layers of various materials that are pressed together to form each plank. A printed film gives the floor the look of real wood or tile and is protected by a tough, durable wear layer that goes on top of the print layer. The inner core is generally made from high-density fiberboard and also forms the tongues and grooves for locking planks together. The core is also the base that all the upper layers and the backing material are fused too. Some manufacturers treat the inner core with melamine resins or water-resistant sealers to help protect the inner core from moisture.

The planks have tongue and grooved edges on all four sides to secure the planks together. Today, most laminate floors use some sort of glueless locking system. Glueless laminate floors can go almost anywhere in the home and are ideal for do-it-yourself projects.

The two main glueless locking systems either involve a tongue and groove that is reinforced from underneath by an aluminum, mechanical locking system or a tongue-and-groove glueless locking system built right into the middle core that allows the planks to snap together during installation.

Some other laminate floors have a tongue that was pre-glued at the factory with special water-resistant glue. Once the tongue is moistened with a wet sponge, it activates the glue and locks the planks together.

There are also laminate floors that require special adhesive to be applied to the tongue and groove at the time of the installation to secure the planks to one another. Professional installation is usually required with this type of product.

Several laminate manufacturers are now offering a microbeveled edging on certain of their styles of plank. Others have added texturing to their surface layer to give their floors a more realistic look.

To learn more about laminate, go to *www.floorfacts.com* and click on laminate floor construction or laminate flooring.

Installing a Laminate Floor

How is a laminate installed?

Laminate floors are meant to be floated over a variety of subfloors and never secured directly to any substrate (subfloor). In other words, the laminate flooring just lies on top of the subfloor, which can be wood, concrete, or an existing floor. This allows the laminate flooring to expand and contract freely as the room humidity levels change. Laminate floors come in planks and tile squares of various sizes and shapes. All four sides of each plank have a tongue and grooved edge for locking them together.

Special cushion is laid down prior to installing laminate flooring. This helps the laminate to float freely over the top of the subfloor, which reduces most sound transmissions. It also restricts moisture from passing into the laminate from the subfloor.

Laminate floors all use the floating floor installation. This means the laminate planks are never secured directly to the subfloor. Instead, the planks are all locked together and float freely over the top of the subfloor. By placing the cartons of laminate in the room for several days prior to installation, you can avoid planks bowing and cupping in the future. The subfloor should be level, or you will have problems getting planks to lock together. Also, planks that are not properly acclimated or subfloors that are uneven can cause the floor to squeak when walked on after installation. Never pound on plank edges during installation or try to force planks together. This will only make fitting planks together even more difficult.

Where should laminate be installed?

Laminate does not belong in wet areas like kitchens, bathrooms, or utility rooms, so do not put it there. Purchase laminates for wood floor areas such as living rooms, dining rooms, dens, hallways, bedrooms, and offices. Measure rooms for laminate like you would normally for a hardwood floor installation, like the examples I have given in the measuring for hardwood section of this book.

What should I look for in a certified laminate installer?

It was only a few years ago when it was required of an individual that wanted to install laminate flooring to take a certification class where he would purchase the tools and gain the experience necessary to start his career in that field. All planks were glued together and straps held them in places until the glue dried, and there are still some laminate products that have to be glued together.

Today this type of certification is not necessary; however it is your responsibility to find out if they are qualified through years of experience and by looking at the installer's references calling them to see what type of job they received from that installer. It is suggested that an installer should have at least three years experience with laminates before hiring them for your job.

If an installer has not installed laminate but has years of experience installing wood plank, he is qualified but not experienced. I would suggest you get an experienced installer to take his place. Installing laminate is not as easy as you might think. It takes experience to do the job properly and finish on time.

Learning about Vinyl Flooring

Today's Vinyl Choices

What is the difference between vinyl and linoleum?

Although linoleum floors have been around for more than a century, many people confuse it with vinyl floors. The fact is the two floors are very different and should not be grouped together.

Linoleum is primarily made of all-natural products, such as linseed oil from flax, wood powder, limestone, resins, and some colored pigments. The backing is made from a natural grass called jute.

There are several manufacturers still producing linoleum floors, such as the Marmorette Collection produced by Armstrong, and the world linoleum leader Forbo has a collection called Marmoleum sheet and tile floors, and finally Domco Tarkett with their Linosom Collection. These companies offer a wide variety of colors and patterns either for commercial or residential use.

Forbo is a European company that has sold linoleum floors in the United States for many years. In commercial applications, linoleum has been used for years because it is extremely durable, burn resistant, has sanitary qualities, and made of all-natural ingredients.

If you are looking for flooring that is different and comes in bright colors, you may want to ask about linoleum floors.

How Vinyl Flooring is Produced

What is vinyl flooring?

Homeowners are offered two types of residential sheet vinyl flooring. The older vinyl floor construction type is called *inlaid* construction, and the newer, more common vinyl flooring construction type is called *rotogravure* construction.

Inlaid flooring is produced by using solid-colored vinyl chips that are laid on top of a carrier sheet and then bonded together with heat and pressure. This process has been around for years, and these floors generally come with geometric type patterns and designs.

Residential inlaid floors have a clear wear layer placed over the top of the chips making the floor's finish easier to maintain. Remember that you are not walking directly on the inlaid chips but you are walking on the clear wear layer that was placed on top of the chips. How long the clear finish will last is dependent on this clear wear layer.

The rotogravure printing process is the most commonly used method for making residential vinyl floors. It also offers unlimited possibilities in pattern and design. This involves a print cylinder that spins around while the vinyl's core layer, which is called the gel coat, passes underneath. The cylinder systematically prints various colored ink dyes to create the pattern. Print dyes are set as a clear wear layer is applied to the surface. The appearance retention of a rotogravure floor is dependent on the durability of the clear wear layer.

The unique manufacturing processes used today can reproduce the look and textures of real ceramic tile, stone, and wood grains. Vinyl floors are manufactured in both sheet and tile. The vinyl sheet floors are offered in both six-foot and twelve-foot widths, and the vinyl tiles come in various sizes and thicknesses. Both types of vinyl floors come in a wide array of patterns and colors.

The more expensive vinyl floors and tiles have better resistance to staining, scratching, gouging, and tearing. Not only is the performance better but so is the warranty on the high-end vinyl floors.

Vinyl tiles are ideal for the do-it-yourself projects in kitchens and bathrooms. They come in both self-stick tiles (adhesive pre-applied at factory) or dry back tiles for glue- down installation. Vinyl tiles range from one-sixteenth-inch thick to one-eighth-inch thick.

More facts about vinyl tile and inlaid sheet flooring can be found in the Internet at *www.wisegeek.com.*

We are going to take a little time to explain the construction and the pros and cons to each type of hard vinyl tile.

What is commercial tile?

The first of the hard tiles and mostly used commercially, and made of vinyl resins is called vinyl commercial tile or VCT. You will find this product in banks, office buildings, hospitals, and shopping malls. It is usually made in twelve-by-twelve-inch tiles and comes in an array of colors and styles. The cost for the materials and installation is reasonable and accounts for its popularity. The drawback to this material is it only comes with a very slight protective coating from the manufacturer. This means that a home or business owner will have to go to the expense of waxing this floor several times before it should be used. Then it must be maintained with periodic waxing for the life of the floor.

What is designer tile?

Alongside of the VCT is a tile made of a vinyl top layer in many different designs like vinyl sheet goods, called "designer tile." It is found mainly in twelve-by-twelve-inch tiles, comes in many styles and colors, and is covered by a thin coat of urethane. However, the edges chip easily, the top can be scratched, and the cost per piece is as high as ceramic tile.

There is another type of designer tile that is made in the shape and design of wood planks. This tile can be placed in wet areas where real wood should not go. The planks of vinyl must be staggered when laid in place like real wood plank.

I have installed both types of tile in the past, and you have to be extremely careful putting the pieces in place, and the floor has to be smooth as glass. It takes a special adhesive, and labor costs about the same as putting in a ceramic tile floor. My personal opinion is that it is not a very cost effective way to go unless you put it in yourself.

Installing a Sheet Vinyl Floor

Sheet vinyl flooring is a practical, durable, and often attractive choice for high-traffic areas such as kitchens and baths. It's also not a big deal to install yourself, if you've got the time and the patience.

How is sheet vinyl flooring installed?

Before you get started, sweep or vacuum the floor to remove all dust and debris. Completely clean the floor as any dirt left behind will interfere with the vinyl adhesive.

Once you have cleaned the area, roll the sheet of vinyl flooring over the floor leaving a little to go up your walls. If your vinyl has squares, lines, or a design, make sure they are even with an outside wall. Also, if your area is a bathroom, make sure the lines or design are even with your tub first before you make it even with a wall.

Now that you have the loose floor in place, place some weight in the center of the room so it won't move around while you are cutting the excess off the walls and doorways. Start with a wall and use a tool to press the flooring into the bottom edge of the wall and cut the excess off with a utility knife. Always pull baseboards and door facing off the walls so the material will be under them, after you are finished installing the sheet vinyl flooring.

After cutting the walls and entries, fold half of the material back over the weight you have in the middle of the floor (this weight could be the 75# roller you rented to roll the floor into the adhesive). Now begin applying the adhesive with the trowel starting with the floor farthest away from the fold and working your way to the fold.

Once you have troweled out the adhesive evenly up to the place where you folded the material, it is time to place the material carefully back where it was before you folded it. After doing so, start using the roller by rolling first along the glue line (where the fold and the adhesive meet) working your way a row at a time away from the fold. This action pushes out air bubbles and presses the material evenly into the adhesive. After doing this, follow the same directions for the other side, troweling and rolling until you get to the outside edges of the room.

Now press the outside edges into the glue with a wide wall putty applicator and cut around the walls. You have now installed a sheet vinyl floor, but remember, it may sound simple but every job is different and how you accomplish the task will change. Doing this job right comes with experience, so I would leave it to the experienced if I were you.

How do I know if I have a good sheet vinyl installer?

Like installers of other flooring products, a sheet vinyl installer should have references you can call before they get started. However, a relationship with your installer needs to begin before they start installing your flooring. Having them come to your home to measure first gives you a chance to meet them. Then you can ask about their references and get a feel for what kind of people they are. If you are not satisfied with them simply call another crew.

An installer of sheet vinyl needs a 75# or 100# roller to roll the floor into the adhesive. There is only one type of floor today that is made of vinyl that does not need to be rolled out, and that product comes with special double-sided tape to tape down the edges only. If the installer does not have a roller and the right adhesive for your floor, do not use them!

Of course, a good installer of any kind of floor should clean up after themselves and are always courteous.

Installing aVinyl Tile Floor

Vinyl flooring, a resilient floor covering similar to linoleum, consists of polyvinyl chloride (PVC) mixed with resin and pressed into thin sheets. Much easier to install than sheet flooring, vinyl flooring also translates to easier repairs and maintenance. When installing this material over cement subfloors, properly prepare the concrete so that the vinyl can adhere to the surface effectively. Follow this same procedure even if you are placing vinyl tile on wood.

What are the steps to installing vinyl tile?

Sweep or vacuum the floor to remove all dust and debris. Completely clean the floor as any dirt left behind will interfere with the vinyl adhesive.

Find the center point of the floor. Measure each wall to find the midpoint then snap a chalk line from each wall to the one opposite; these lines will intersect at the center point. Start your installation at this point, which will improve the chance that the floor will appear straight and even, regardless of whether the walls are straight or crooked.

Lay out your tiles. Perform this step without adhesive so you can decide on patterns and layout. When you've finished, stack the tiles in the order you plan to install them. You may wish to have a stack for each row if you plan to use a complex pattern.

Spread vinyl adhesive over half of the floor and allow it to harden slightly before installing your tiles. Press the tiles into the adhesive starting at the center of the floor, butting each tile tightly to the adjacent units. Cut tiles with a vinyl cutter or utility knife as needed.

Repeat step 4 to complete the other half of the room. If using peel-and-stick tiles, skip the adhesive and simply peel the paper backing from each tile before sticking it in place. Roll the entire surface with a one-hundred-pound floor roller when finished to ensure a firm installation.

To find out more about do-it-yourself installations, google "Installation of Flooring" and then pick the type of floor you want to install. You can also go to *www.ehow.com* for more information.

How can I tell if I have a good installer of vinyl tile?

Most vinyl tile installers have been installing for many years, especially if they have been doing large projects like hospitals or schools and such.

Look for someone that has installed thousands of square feet of tile in his career and you will have a good vinyl tile installer.

LEARNING ABOUT HARD TILE

Today's Hard Tile Choices

In this section, we will learn about the different types of tile I call *hard tile*. Hard tile is any tile made primarily of natural ingredients or which is found in nature. For example, ceramic tile is made primarily of clay, and granite is taken from the mountains in its natural state to make granite tile. We will discuss each of these tile choices in detail.

What is ceramic tile?

> Ceramic tile is a natural product made up of clay, a number of other naturally occurring minerals, and water. Ceramic tile comes in both glazed and unglazed tiles. Glazed ceramic tile has a ceramic coating applied to the tile body that gives the tile its color and finish. Ceramic tile is a popular choice for your interior floors and walls.

Some characteristics of glazed ceramic tile are the following:

- **Durable**—a properly installed ceramic tile will outperform and outlast nearly any other floor covering product created for the same application.
- **Easy care**—glazed ceramic tile resists stains, odors, and dirt and can be cleaned up with a damp mop or sponge or common household cleaners.
- **Scratch resistant**—Grade III and Grade IV glazed ceramic tiles are extremely resistant to scratching like you do with some floors.

- **Environmentally friendly**—ceramic tile is manufactured using natural materials and does not retain odors, allergens, or bacteria.
- **Beautiful and versatile**—modern ceramic manufacturing technology has created virtually an unlimited number of colors, sizes, styles, shapes, and textures that will add beauty and character to any room.
- **Fire resistant**—ceramic tile doesn't burn nor emit toxic fumes. A lighted cigarette, when dropped on the floor, or even hot kitchen pans or skillets will not scorch or melt the surface of glazed ceramic tile.
- **Water resistant**—most glazed ceramic tile has a dense body that permits little or no accumulation of moisture. Some of the other important things to consider when selecting a ceramic tile are as follows:

  - Slip-resistance of ceramic tile
  - Size of tile compared to the overall room size
  - Width and color of the grout joints
  - Thickness of the tile
  - Height variations of the floors between adjoining rooms
  - Suitability of the subfloor for ceramic tile
  - Cleanability of the ceramic tile

What is porcelain tile?

Porcelain tiles are growing in popularity with homeowners and interior designers. They are denser and less porous than glazed ceramic tile and are highly resistant to moisture, stains, bacteria, odors, and even harsh cleaners. Porcelain tile is especially resistant to staining, scratches, and fading. This is probably why our sinks, bathtubs, and toilets are made of this material. They are available in both polished and matte finishes and in a variety of sizes, colors, and textures.

Porcelain tiles come in either glazed porcelain or a through-body porcelain tile. Through-body porcelain tiles have the same colors all the way through from the front to the back of the tile, so if they are chipped or scratched, the color will not change. However glazed porcelain tiles are similar to glazed ceramic tiles in that they have a design layer (glaze) on top of the tile body. The tile body, which is a different color, will be noticeable if chipped.

True porcelain tiles are freeze-thaw stable. This is why porcelain tiles can be used outdoors in exterior settings, as well as indoors. Porcelain tiles are formed under extremely high pressure and fired at very high temperatures. This makes these tiles much denser and stronger than the common glazed ceramic tiles, so they are ideal for entryways, corridors, and other high-traffic areas. Also, porcelain tile requires special setting materials for bonding to the subfloor.

What is travertine tile?

Travertine tile comes from a naturally occurring stone that has a natural rich beauty that has been adored and used as a building material since the Roman Empire. These stone tiles have unique veining and a subtle blend of beige color variations caused by the sedimentary rocks, minerals, and limestone deposits that crystallize over hundreds of years. Travertine tiles are characterized by their irregular edges and surfaces.

These tiles can be used on both the floor and wall in both indoor and exterior applications. They come in filled (pores and holes filled) and unfilled stone tiles, as well as polished or tumbled tiles. Each tile is unique in both variations of color and in general surface appearance and veining. Tumbling the travertine tiles gives the tiles a more rounded and timeworn or ancient appearance.

What is marble tile?

Marble tile is a timeless fashion classic that has been used for many centuries as a building material as well as for sculptures and monuments. Marble consists of sediments like seashells or other ocean debris. Heat and pressure over time eventually crystallize these ingredients into marble. As the marble stone forms, fissures filled with minerals create the beautiful veining colorations that give this stone its visual appeal. This also makes each marble tile unique and never exactly alike. Polishing the tiles further enhances the visual appearance and distinctive colorations within each tile.

Marble tiles are usually only used in interiors as floor tile or around fireplaces and in bathroom shower or tub enclosures. These tiles come in many different sizes and various natural earth-tone colorations.

What is natural stone tile?

Natural stone floors offer a distinct and beautiful alternative to your flooring choices. Each stone has its own unique visual appearance. Unlike glazed ceramic tiles, stone tiles do not have a protective glazed coating on top, and the color goes all the way through the tiles.

Slate is a fine-grained rock with traces of metal that were present during its slow, natural formation in the earth. Natural shade variations are an inherent characteristic of slate and enhance the distinctive details of each piece. Slate has a rustic charm that appeals to our senses for natural materials.

Granite is a natural stone and is one of the hardest and most durable of all stones used in flooring. The speckled colorations found in granite a beautiful and subtle fine pattern of color. Polishing the granite adds a visible sheen and depth to each tile. Like marble, no two pieces are alike, and color variations add to the natural beauty and appeal. Professional installation is highly recommended for installing granite floor tiles.

Installing a Hard Tile Floor

Can you install tile directly to wood surfaces?

> Chip board, cushioned vinyl flooring, particle boards of
> any type, Luann plywood, OSB (Oriented Strand Board),
> tongue and groove planking, and hardwood floors are
> unsuitable substrates to directly install ceramic tile over.

> Although it can be done successfully, many experts believe
> that ceramic tile installed directly to plywood surfaces
> should be avoided whenever possible. Plywood has a smooth
> surface and tends to swell, warp, and delaminate when it is
> exposed to moisture. Install at your own risk.

Can you install tile over vinyl or linoleum floor coverings?

> Installing ceramic tile directly to vinyl or linoleum surfaces
> should be avoided whenever possible. Install at your own
> risk.

> If you are going to install over this type of floor, the vinyl
> or linoleum flooring must be a noncushioned type and
> securely attached to the subfloor. If the floor covering does
> not contain asbestos fibers, we recommend that the surface
> be scarified or sanded to provide a rougher surface for the
> thin set mortar to bond to.

> Install ceramic tile using a latex modified thin set mortar
> approved by the manufacturer for installation over vinyl
> and linoleum surfaces.

Should you install tile over ceramic tile backer boards?

> Cement ceramic tile backer boards may be installed
> over plywood subfloors and should be secured using
> one-to-one-fourth-inch, corrosion-resistant roofing nails
> or one-to-one-fourth-inch ribbed wafer head screws in

combination with a thin set mortar bed. Screws or nails should be installed every six inches to eight inches on center.

Backer boards will add to the height of your new floor and may require height reducing thresholds or transition strips where tile meets carpet, vinyl, etc. Doors may also need to be trimmed. Refer to the backer board manufacturer for specific product recommendations and limitations.

To prevent moisture from damaging the plywood subfloor, we recommend that a waterproofing membrane be applied over all ceramic tile backer board surfaces when installing in wet areas like shower and tub wall facings.

Can you install tile directly to concrete slabs?

Paint, cutback adhesives, gypsum-based fillers or levelers, sealers, or chemically treated cement slabs are unsuitable surfaces to install ceramic tile over and should be removed by nonchemical methods whenever possible.

Concrete slabs must be thoroughly cleaned prior to the installation of tile. To remove dust, mop cement slab using clean water only and allow drying completely. Very smooth concrete may be roughened up or etched using an acid-based solution designed for this purpose.

Make sure to fill in and float off any dips, humps, or waves on the concrete foundation using a Portland cement-based floor leveler. For dips, this product may be used to fill the cavity and screed off using a level or straight edge. For humps, apply the floor leveler around the base of the protrusion. This will help to lessen the impact the protrusion will have on your finished floor.

Most Portland cement-based floor levelers need to cure for at least twenty-four hours before the tile can be installed.

Learning more about natural tiles and their installation is easy, just go to "Facts about installing ceramic tile" and follow the prompts.

How can you tell you have a good hard tile installer?

If an installer comes to your home and says he installs every kind of hard floor surface, look in the back of their truck because the proof is in the tools they carry to the job. If they have a tile cutter, a wet saw, a mud mixer, and bags of thin set and grout, you can bet they know what they are doing because they have thousands of dollars worth of tools with them, which means they have done a lot of work and are serious about their trade.

I would however get some references and call them before they go to work because there are some out there that have plenty of tools and but have done a lot of bad work, and it has not caught up with them yet.

Be careful out there when hiring installers separate from the retail flooring stores. Those stores have installers that have proven themselves able to do the jobs they are assigned, and that is why they continue to use them. The installers that could not do the job right were let go. Those may be the very installers you find!

MEASURE YOUR SPACE

The ability to accurately measure and estimate your flooring need is especially important when ordering your own materials. This section will guide you on how to measure rooms properly, accurately identify where seams or breaks should be placed (if needed), and give you an idea of the total cost of materials for your room or rooms.

Measuring for Carpet

Now that you have picked out just the right carpet and pad or you have decided to glue the carpet down, the next step is measuring the area. With the proper measurements, a flooring salesperson can give you an estimate of cost, even over the phone. However, most carpet stores in your area will insist they measure your floor for accuracy before ordering your floor. Even a wholesale carpet outlet online will want you to fax them a diagram of your job to make sure you are ordering enough material.

What do I do first?

The easiest way to measure for carpet is to draw out the room or rooms on a piece of paper. If the rooms connect (example: a living room, den, and hallway are all connected together), you must draw out the rooms as they lay together. For your convenience, we have placed graph paper in the notes section in the back of this book, so you can draw out your area and put measurements there. Carpets must be run in the same direction, especially where rooms connect. The nap of most carpets lay in one direction. When carpet is made, the yarn is sewed into the backing at an angle, and this angle is called nap (pile angle). The nap needs to lay in the same direction when seamed together in larger rooms or when the same carpet is laid in multiple rooms that touch each other. It is easy to tell where the carpet is put together when carpet is turned the wrong way.

Next, you must measure long point to long point (both the length and the width), measuring into doorways to get proper square footage or yardage. It is also helpful to measure where the walls change in the room. For example, say you have a hallway that changes from four foot wide to six foot. Measure the wall all the way to the change then follow the wall measuring each section of wall in your drawing, placing the measurement on that section of wall. Also make sure the length and width figures of that hallway (or any room) are documented where those changes occur. *Take a look at the example diagram I have placed at the beginning of the graph paper in the back of this book.*

Without these measurements, you or the salesperson will not be able to figure your flooring needs properly. Remember to put exact measurements on your diagram and let the salesperson or installer know that your measurements are exact. They will add a few inches to each length and each width for waste if necessary.

How do I figure how much carpet I need?

Today most carpeting comes twelve feet wide. Others will be thirteen feet six inches and also fifteen feet. You can buy carpet in almost any length, but the width is determined by the manufacturer. So when you look at a room, you have to remember that the carpet is normally twelve feet wide. For example, say you have a room that measures ten feet four inches by thirteen feet three inches exactly. If the carpet is made twelve feet in width, then you will need a piece for that room that is thirteen feet six inches long. (Installers always add three inches from exact measure, so a little carpet can go up the wall.) The difference between ten feet four inches and the twelve feet width of the carpet is waste.

Now to figure square footage including waste for this example, you multiply twelve feet by thirteen feet six inches (12 x 13.5). You must now divide the square footage (162 square feet) by nine if you want the right figure for square yardage (18 square yards). The retail flooring stores of today deal mostly in square footage.

How do I know where the seams should go?

Dealing with multiple rooms or extra large rooms is more complicated. Allow your sales professional or carpet layer to figure multiple rooms or large rooms. Remember to tell them you want as few seams as possible. However, if you don't want to involve someone else in the measuring process, I will help you with an example that is a little hard to explain.

Let's say you have a living room that measures twenty-four feet three inches by fifteen feet eight inches. The carpet you selected comes in twelve feet widths. There is a door leading into the kitchen on the wall running the twenty-four-feet-three-inch length and another on the opposite side going into a hallway. Both of these doors are at the far end of the room against an outside wall that has windows, and the distance to the center of each door is fifteen feet eight inches.

Usually we do not want the seam to cross a high-traffic area. However, in this example, the best place for the seam will be closest to the door that leads into the kitchen. It runs the full length of the room twenty-four feet three inches plus three inches up the walls (total 24'6"). A twelve feet by twenty-four-feet-six-inches drop will be placed against the wall of the hallway door, and a four-feet-wide piece will run the length of the room (24'6"). To fill the balance of the room, a twelve-feet-wide piece of carpet that is eight-feet-six-inches long will be cut into three pieces and each piece will be four feet by eight feet six inches.

They will then be seamed together to fill the length of the room. These three pieces (4 x 8'6") will be seamed together end to end (running the nap the same direction) and the side of those carpet pieces (which is now 25'6" long) will be seamed to the edge of the twelve feet by twenty-four-foot-six-inch drop.

After the carpet is seamed together (total size of the seamed pieces is now 16' x 24'6"), it is time to stretch in the carpet. (I told you it was hard to explain!)

There are so many things to consider when dealing with seams. Room size, door placement, outside windows, and size of the room all have to be considered when figuring where seams should go. It would be hard to tell you how to figure where the seams should be placed in every situation. It comes with experience. However, an example of a whole house installation and where seams should be placed is in the back of this book along with several sheets of graph paper to draw out your own rooms.

Can I estimate how much flooring I need without knowing where the seams should go?

If you want a close square footage estimate on materials without knowing where the seams go, figure long points (into doorways) in each room and multiply the length and width together. Then add those square footage figures (per room) together and divide by nine (if you want square yardage), then add 10 percent waste (square yardage or footage x 1.10). You should be pretty close to the material needed for your job.

How do I estimate material cost?

Now that you know how many square feet/yards of material you will need, you should be able to estimate the cost of materials by multiplying cost of each item (per foot or yard) by the total feet or yards of material needed.

For example: Let's say you have a room that is 10.5 x 14.75, which means you will need a piece 12 x 15 (Carpet comes twelve-feet wide, and the installer needs three inches up the wall for waste).The carpet you chose runs $1.99 per square feet, and the pad runs $.67 per square feet. Now add those two together ($1.99 + $.67) then multiply times the number of square feet (180) of material you need for that room. The total will be the total cost of these items.

$1.99 + .67 = $2.66 x 180(12 x 15) = $478.80 plus tax.

In this example, $478.80 plus tax will be the total cost of carpet and padding.

Measuring for Hardwood

Now that you have decided to define your home or office with the rich look of wood, it is time to measure the area. This type of flooring is measured in square feet and always comes in cartons. The cost of installation by a certified technician is much more than carpet, but the results can be seen for years to come.

How do I measure for a wood floor?

When measuring for wood, you must get as close as you can to the exact square footage of your floor space and then add 10 percent. For example, you have a room that is fourteen feet five inches wide and eighteen feet two inches long at the longest points. You also have an entryway that connects to that room which is exactly six feet by nine feet. I would multiply fifteen by eighteen and add six by nine to it, and then add 10 percent waste.

(W x L + W x L = square footage x 10% or 1.1 = square foot need for 2 rooms)

It is usually required, when buying hardwood flooring, that you must purchase the next full box. For example, say you need 560 square feet, which includes your 10 percent waste. If the type of hardwood you purchased comes 30 square feet to the carton, you will end up buying nineteen cartons or 570 square feet total.

Do hardwoods come in different grades like lumber?

Hardwoods today are graded, and their price reflects it. When shopping for wood for the home, you will hear such terms as cabin grade or builders' grade. The higher the grade, the less culling or searching for acceptable pieces you will do. In the lowest grade, there can be as much as 25 percent unsatisfactory pieces to every carton. Warning: Be careful what you buy, and remember, your installer will charge you for every square foot he culls.

Measuring for Laminate Flooring

There are some advantages to installing laminate floors over wood floors. For instance, wood can dent or scratch more easily than laminate, and wood generally costs more. There is also a cost of maintaining some wood floors, and laminates are easier to install yourself. If that makes sense, you are probably ready to measure your area for laminate.

How do I measure an area for laminate?

When a laminate is being installed, the area is measured the same way as if you were installing a hardwood. You will be purchasing laminate by the square foot and also by the carton like a hardwood. Also, the figuring method is the same. However, there is very little culling (picking through for best pieces), if any, when dealing with a laminate. Go back to the subheading "Measuring for Hardwood" and follow the directions there if you need more guidance.

Will I need to figure in extra for warped or damaged flooring?

Most laminate planks, unless damaged while being transported, are ready for installation. Usually 5 percent waste or (square feet x 1.05) will be sufficient in your figuring of square footage. However, order a little extra (maybe a carton or two) just in case there is damage now or in the future. If you accidentally damage your flooring in the future, that damage can be replaced with a new piece, if it is available.

Measuring for Sheet Vinyl

If you are looking for an inexpensive way to cover a bathroom, utility area, and a kitchen or entry, sheet vinyl may be the best choice. The heading "Measuring for Sheet Vinyl" includes measuring for linoleum, inlaid, rotogravure vinyl flooring, and the newer types of sheet vinyl now available. Sheet vinyl flooring normally comes in six and twelve feet wide sheets and can be ordered as long as you need (usually up to around 150 feet), depending on the type of vinyl and the manufacturer.

How do I measure for vinyl flooring?

Measuring a room for vinyl is like measuring the room for carpet. Go back to the subheading "Measuring for Carpet" and follow the directions there if you need more guidance. However, when the room is wider than the material's maximum width, a seam is necessary (just like carpet). If the vinyl has a pattern, it means a pattern match has to be considered (in sheet vinyl as well as in a carpet that has a pattern). Usually a 10 percent waste should be considered when figuring the square footage material need of a room or rooms if you have seams in vinyl. And even more waste should be expected when your flooring has a pattern.

How do I figure for seams in sheet flooring?

Figuring the total material need (example: square footage plus 10 percent waste), when a room is larger than the material is only a rough estimate.

In order to know exactly how much material you need, a person would have to know where to place seams, the pattern match (6", 9", 12", 18", 36", etc.), and how to lay out the cuts of material properly. This will take the expertise of an installer or someone from a flooring store that gives estimates.

I will however give you an example to show you its complexity.

Let's say you have a kitchen that is eighteen feet three inches in length and thirteen feet nine inches in width. You have selected a ceramic tile pattern that has a nine-inch pattern match. In other words, the pattern repeats itself every nine inches, and it measures from center of grout to the center of grout both ways. This vinyl is manufactured in twelve-foot widths, so the kitchen will have a drop of material which is twelve feet by eighteen feet six inches and a fill of one foot nine inches by eighteen feet six inches.

Now it's time to figure how much material you will need for that fill with pattern match. The space needing filled is one foot nine inches; however the pattern is nine inches. To go past the exact fill amount (1'9") and make the best use of the extra material, you have to end your cuts on nine inch centers. In other words, the fill width will now be figured at twenty-seven inches (9 x 3 = 27) or two feet three inches, instead of one foot nine inches or twenty-one inches.

Now how many pieces two feet three inches wide can you cut out of a twelve-feet-wide piece of material? The answer is five. So now we can cut five pieces from that twelve-feet-wide vinyl, but how long will that piece need to be to fill an area that is eighteen-feet-six-inches long?

Well, you take eighteen feet six inches and divide it by five (about 3'9"), and that will be the exact fill amount, but remember the pattern goes both length and width. How many nine-inch ceramic tiles with grout will go into three feet nine inches? The answer is exactly five. And what is five by three feet nine inches? The answer is eighteen feet nine inches.

Looks like you will have enough for the fill if you order an eighteen feet six inches plus three feet nine inches for the fill, which is a total order of twelve feet by twenty-two feet three inches. However, I would order a little extra if I were you.

That was one example of the many possibilities that are out there. It's not as easy as it sounds. Before you order, make sure you get a second opinion.

Measuring for Vinyl Tile

Just like ceramic or porcelain tile, vinyl tile is manufactured in many different grades, sizes, colors, and thicknesses. Measuring for a vinyl tile installation is much like figuring the square footage of any tile, even carpet tile. You purchase the tile by the carton, and the carton size or the number of square feet to a carton is figured by the manufacturer.

How do I measure my room for vinyl tile?

You must measure long point to long point both width and length and multiply the width and length together in order to figure square footage. This total is the base square footage (the square footage without waste) of that room.

Once you have the base footage figure you must add 10 percent, and because this type of tile is laid tightly against one another in a special adhesive, there are more tiles per square foot than with ceramic or any natural tile. Thus more pieces per square foot will be purchased for vinyl than for natural tile. This means you will need to buy more vinyl tile than natural tile.

Measuring for Hard Tile

Naturally made tile is cut into many sizes, shapes, and thicknesses. There are so many hard tiles to choose from, but once you have chosen to do a room or rooms in this type of flooring, you are really upgrading your home or business. This choice tells something about you. It will tell people you are there to stay because it is not something you can easily pull off the floor when you are tired of it.

How do I measure for any tile accurately?

Hard tile (ceramic, porcelain, marble, granite, etc.) installation areas are measured by the square foot. You can put hard tile almost anywhere, even outside in the weather if you have the right tile.

It is better to let a salesperson or installer estimate your tile needs on walls, showers, backsplashes, or countertops. However, if you are installing ceramic, porcelain, or any other hard tile on floors, use the same directions for measuring hardwood or vinyl tile. If you are installing it yourself the rule is, long point to long point, multiply, and add 10 percent for each room. This will give you a *rough estimate* of what your total tile square footage with waste will be. You can then figure your square footage into number of cartons needed, depending on your tile choice. It is not a bad thing to have a few tiles left over, so go ahead and order plenty.

Is there anything else I need to know when considering any type of hard tile?

Along with the tile, there is another choice that needs to be addressed: grout. Tiles are not installed next to each other like vinyl tile. A gap between tiles is set by the installer. Grout fills in the gaps between tiles and is used to accent the room. Grout comes in many different colors and consistencies and is used whenever ceramic, porcelain, marble, slate, and any other natural tile is being installed. Bags of grout must be considered when figuring cost and the amount of tile needed for your job.

Remember that hard tiles can go anywhere in the house or office. The tiles themselves are not restricted by moisture or high traffic like other flooring products. Only the subfloor is affected by moisture. If you feel a tile floor is cold and hard, you're right! If you are having a problem with that, it is easily resolved with an area rug.

These floors are nearly maintenance free, and because of today's time restraints, they can add to the value of your home. The bottom line is hard tiles are an investment. Initially they may be more expensive, but in the long run, it will save you money.

Let's Go Shopping

Searching for the Right Place

Today there are many choices when looking for the right place to shop for carpet, hardwood, tile, or any other type of flooring. Let me tell you what I would do if I were ready to buy flooring.

Where should I go to buy flooring?

First, I would talk to my friends or family and ask them if they have had flooring installed recently (last five years). I would then ask where they went to buy the flooring. Next I would ask them if they were treated well and if the installation was done to their satisfaction. Finally, I would ask them if they thought the price was fair and if they would go back when they needed more flooring. If you get a yes from all these questions, I would begin there.

Second, if I did not have friends or family that had flooring done recently, I would then look up flooring stores (in the Internet or in the yellow pages) that are in my area (within a fifty-mile radius). Remember, a good flooring store located near you is better than ordering carpet direct from the manufacturer and finding your own installer or installing it yourself.

Why should I consider buying my flooring from a local retail store?

The reason for this is simple: if you buy from a local flooring store and if there are any problems with the flooring, installer, or installation, you can go right back to that store and get the problem resolved. It is much more difficult to get problems resolved when the flooring is dropped at your door and you found the installer in the yellow pages.

However, it does not hurt to know the cost of flooring that can be purchased online, even if you have decided to buy from a local flooring store. It will keep the store from overcharging you when it is time to purchase the installation package because you will know the general wholesale price of that item.

If you have decided to order direct and do-it-yourself or hire an installer, the first thing I would do, if you have access to the Internet, is google "flooring stores online" or go to *www.carpetbuyershandbook.com.*

By accessing flooring stores online, you will see pages of stores online you can order your flooring from. Do some comparison shopping to get the best price, but be careful when ordering online. You must pay for the whole order before they will ship the merchandize to you, so make sure the company you are ordering from is legitimate.

What are some reasons why I should not buy from a local store?

Do not buy from a retail store until you are sure they are a quality company and will stand behind the work of their installers.

Products and services from local flooring stores will most likely cost more than buying direct because of overhead.

Other than these two reasons, I see only advantages with buying your flooring from a local retail store.

What are some reasons why I should not buy direct in the internet?

- After ordering, you will need a safe dry place, with a temperature between fifty-five and seventy-five degrees F to store the materials until they are installed.

- If the materials are delivered damaged, it is difficult and time consuming to get those goods replaced.
- Warrantees on flooring are void if a certified installer does not install that flooring.
- You have no one but your installer to back up the installation guarantee.

Knowing What to Bring

Now that you have decided on a place you might buy your flooring, you will need several items as you get ready to shop for flooring. I will explain each item and why you need each one.

Here is the list:

1.) You need a drawing of your room or rooms, (whatever you want covered by flooring), with exact measurements of lengths, widths, wall lengths, closets, stairs, and where doorways and windows are in each room. (Without measurements they cannot give you a total price, and you will have to wait for the salesperson to come to your home where they hope to sell to you right then.)

2.) You need an idea of what type of flooring you want in each room. Remember to go by the suggestions in this book and be careful of wood products in wet areas. (If you do not know what you want, the salesperson will sell you what they think is right.)

3.) You need a pillow sham, paint swatch, or anything that will help you match the colors in your home with the flooring at the store. (Anything that will help you color coordinate your choice at the store because you will waste a lot of your time taking samples back and forth just to get the right one.)

4.) You need plenty of time to find just the right floor for each area. (If you are in a hurry, you will always wonder if you made the right decision. Remember, your new floor will be there many years to come, so take time to pick the right one.)

Making Contact with the Store

When you get to the store, you will be approached by a person who wants to make a sale. Most salespeople work on some kind of commission. If you know this, you will understand that time is money to them, and it will be their desire to hurry the sales experience along.

After reading this book, you will be more prepared than any other customers they will meet. You will have control over the whole sales experience. They will ask if you have measurements to the rooms you want to cover, and you will say yes. They will be surprised to see you enter with a paint swatch or a pillow sham to make sure the color of the flooring is right. And you will have an idea of what flooring you want to buy and where it should go in your home. And you will pay a reasonable price for the floor you purchase.

By having all of this, you will have complete control over the purchase of your flooring and plenty of time to find just the right one.

What happens after I have picked out all the flooring I want installed?

After you have picked just the right floor and you feel confident about your decision, it is your right to get an estimate from the salesperson. There are two ways to get an estimate.

One, get an estimate on what it would cost if you let them do everything. Move the furniture (if there is any), take up old flooring, and haul away, floor preparations and installation. Then all you have to do is write a check. This is the way most people do it.

Two, you can ask for an estimate on just the cost of materials. Even if you do not have an installer in mind, just get the "cash and carry" price. It makes the retailer feel like a wholesaler, and he most likely will give you a special price for materials only. If you do not find an installer on your own, then the retailer will probably suggest one. The installer will most likely negotiate a price, and you could save money on the installation. Even the store will love you because they will not have to worry about how or when it will be installed.

After agreeing on a price, whether it is materials only or an installation package, you will need to put down a deposit in most cases. Usually, half of the total cost of your purchase is required to order what you have selected. The balance will be due when materials are picked up or after the installation is completed.

After the installation of your floor, the retail store normally places a one-year guarantee on that installation. Even if the original installer is not available, the store will send out another competent crew to fix the problem if one should arise. This is another reason why using a local store can be better than ordering your materials online and hiring an installation crew yourself.

CONCLUSION

It is my sincere desire that the information in this book, which came from decades of experience, might be used to save you time, money, and the frustration that comes from receiving goods and services that are not acceptable. I have seen this frustration many times in the eyes of those I have helped repair problems that another installer would not or could not fix. I have also seen the anger in a homeowner's eyes when a floor was sold to them that did not hold up as promised.

By using the tips I have recommended, you will know what type of floor is the best for the areas you need covered, you will have a drawing with measurements of those rooms, and you will have the color that will best match each room. Now all the salesperson needs to do is direct you to that flooring you have said you wanted to see, guide you to their best sale items, put some figures together from your drawing, and make sure you have their best installer scheduled for your job. It is their job to make sure you are happy, start to finish. You're the customer; you should get what you pay for!

In closing, I hope that you have learned something about the flooring industry (products and installation). My purpose in producing this book is to help the flooring shopper assist the "true flooring sales professional" in their job so that both parties will save time in the shopping experience. Take this book whenever you go to buy flooring. You'll be glad you did!

Remember, this guide was developed so you can take it with you when you shop. It has a place in the back where you can take notes. It also has a place where you can draw out your room, put your measurements, and place the name of the flooring where you want it. I tried to make it as easy as possible. Life is too short to be worry about whether you are prepared to go shopping for flooring!

Special credit goes to the following:

Mohawk University Workbook, Mohawk Industries Inc.
Floorfacts.com
Facts about Installing Ceramic Tile
Floor Buyer's Guide
Floorstransformed.com
Ehow.com
Wisegeek.com

FLOORING MEASUREMENTS SECTION

This section was developed so you might have a place to draw out the rooms of your house like a professional; placing the name of the room, where the doorways are, the length and width of the room, and what type of floor you want installed there.

On the next page you will find an example of rooms drawn out on graph paper. The squares in the graph paper can represent one square foot each or even two squares per foot if you want. In the example it is two squares per foot. There are a total of 72 squares in length and 40 squares across in width on each graph. I placed measurements and the name of the rooms, plus the type of flooring I wanted to put in those two rooms on the graph.

On the back side of each of the six graphs there is a place you can take notes, in case all your ideas will not fit on the graph.

A tape measure and a pencil is all you will need to place your floor plans on the graph. Remember to measure halfway into the doorways and place exact measurements with an arrow pointing in the direction you measured that part of the room. You will see what I mean when you study the example graph.

I hope this added feature helps you feel more prepared as you go out to shop for flooring.

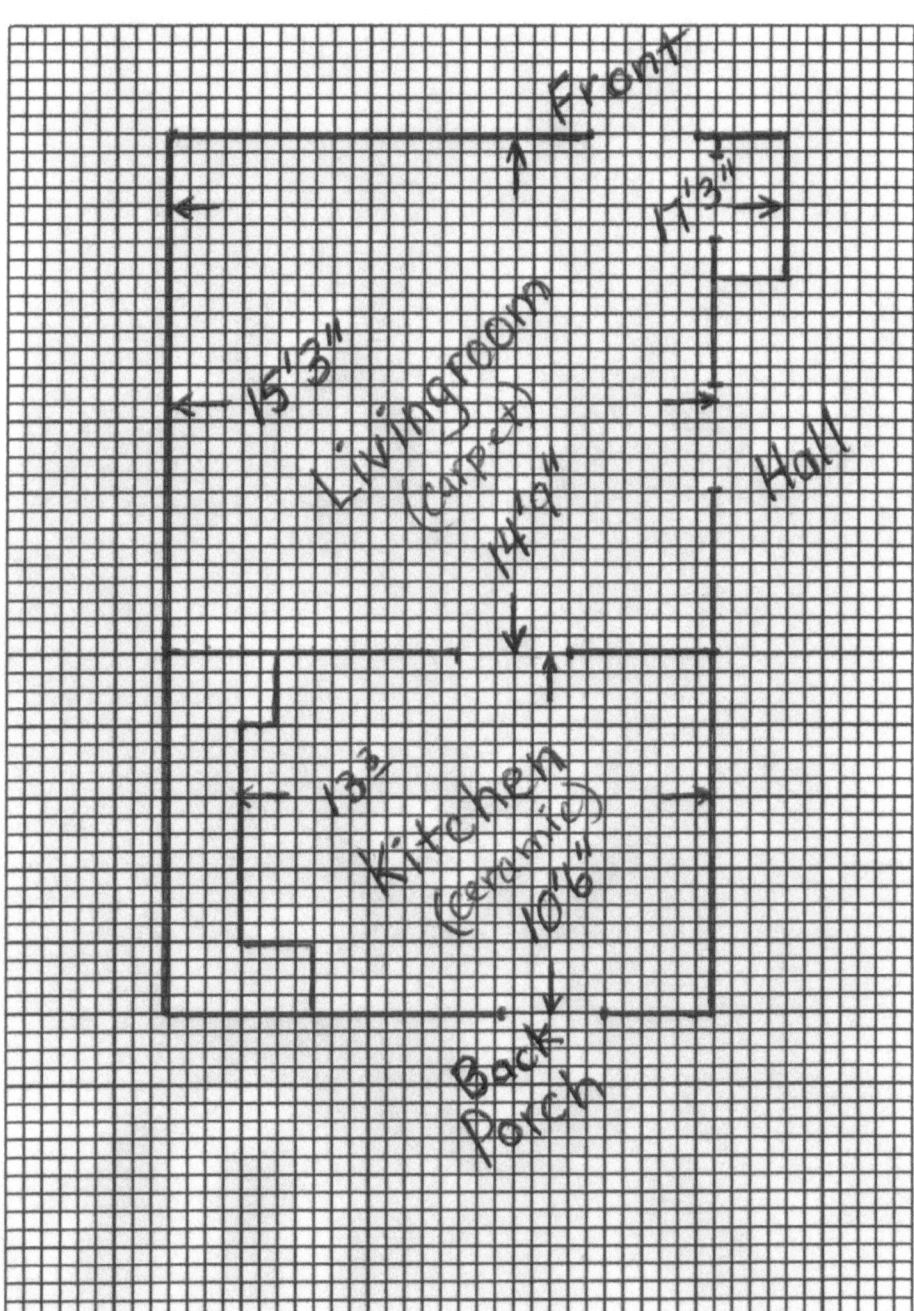

NOTES

NOTES

NOTES

NOTES

NOTES

NOTES